AI: The Hype, The Reality, and The Future
By Anas Bedraoui

Copyright © 2024 by Anas Bedraoui. All rights reserved.

No part of this book may be reproduced, distributed, or transmitted in any form or by any means, including photocopying, recording, or other electronic or mechanical methods, without the prior written permission of the publisher, except in the case of brief quotations embodied in critical reviews and certain other noncommercial uses permitted by copyright law.

ISBN: 9798328698832

Disclaimer:
This book is a work of non-fiction. While the author has made every effort to ensure the accuracy and completeness of the information contained herein, the author assumes no responsibility for errors, omissions, or changes to the information contained in this book. The reader should consult with a professional where appropriate.

First Edition: June 2024

Introduction

Artificial intelligence (AI) is one of the most discussed and impactful technologies of our time. From self-driving cars and voice-activated assistants to advanced medical diagnostics and financial forecasting, AI's influence seems limitless. However, alongside these impressive advancements lies a world filled with misconceptions, exaggerated claims, and what we might call "Artificial Illusion."

This book peels back the layers of hype to provide a clear perspective on AI. Tracing the history of AI from its inception in the 1950s to its current state, we will delve into the cycles of enthusiasm and disappointment that have marked its development. We will examine genuine breakthroughs that have significantly impacted industries and improved lives, as well as superficial applications that have inflated expectations and sometimes misled stakeholders.

Distinguishing real innovation from hype is crucial. As AI integrates further into our daily lives, it is vital to ask the right questions, demand transparency, and critically assess the claims made about this powerful technology. This book offers the tools and insights needed to navigate the AI world with a balanced and informed perspective.

Throughout these pages, we will highlight the unique qualities that set human intelligence apart from artificial intelligence. Human creativity, emotional intelligence, and critical thinking are irreplaceable assets that AI, for all its capabilities, cannot replicate. Appreciating these human elements helps us understand how to leverage AI to enhance our abilities rather than replace them.

Ethical considerations surrounding AI form another critical aspect of our examination. Addressing issues of bias, transparency, privacy, and accountability is essential to ensure that AI is developed and used responsibly. Through case studies and real-world examples, we will illustrate the importance of ethical guidelines and standards in fostering trust and promoting fairness.

Looking to the future, the potential for AI to revolutionize various fields is immense. Realizing this potential requires a commitment to continuous learning, ethical development, and genuine innovation. This book provides a guide for navigating the complexities of AI, emphasizing the importance of maintaining a balanced perspective and fostering collaboration between humans and machines.

In a world increasingly driven by technological advancements, remaining grounded in the principles of honest work and true innovation is essential. Understanding AI's past, appreciating its current capabilities, and critically assessing its potential allows us to harness its power to create a better, more equitable future for all.

Welcome to a comprehensive look at AI—its history, its realities, and its future. Let us approach this with curiosity, skepticism, and a commitment to uncovering the truth behind the technology that promises to reshape our world.

Chapter 1: The Hypes of the 1950s and 1960s

Early Technological Advances and the Dawn of the Computer Age

The 1950s heralded the dawn of the computer age, an era that saw the birth of digital computing machines capable of performing complex calculations at unprecedented speeds. These early computers, though rudimentary compared to today's standards, represented a monumental leap in technology. The ENIAC (Electronic Numerical Integrator and Computer), completed in 1945, is often considered the first general-purpose electronic digital computer. ENIAC could solve a large class of numerical problems through reprogramming, setting the stage for future developments in computing.

These early machines sparked the imagination of scientists and engineers. Theoretical work by pioneers such as Alan Turing and John von Neumann laid the foundational principles of computer science. Turing's 1950 paper "Computing Machinery and

Intelligence" introduced the concept of the Turing Test, a measure of a machine's ability to exhibit intelligent behavior indistinguishable from that of a human. Turing's question, "Can machines think?" opened up new avenues of exploration and debate about the potential of artificial intelligence.

Public Perception and Media Portrayal of AI and Automation

The burgeoning field of AI did not exist in a vacuum; it was heavily influenced by public perception and media portrayal. The 1950s and 1960s were times of rapid technological advancement and significant social change, and the idea of intelligent machines resonated deeply with the public. Science fiction, a genre that had always flirted with the concept of artificial beings, began to explore AI more explicitly.

Isaac Asimov, a prolific science fiction writer, was among the first to popularize the concept of robots with his collection of stories "I, Robot," published in 1950. Asimov introduced the now-famous Three

Laws of Robotics, which governed the behavior of robots and ensured they would be safe and beneficial to humans. These stories, while fictional, influenced public expectations and helped shape the cultural understanding of AI.

Television and movies also played a significant role in shaping public perception. The television series "The Twilight Zone," which aired from 1959 to 1964, featured several episodes that explored the implications of advanced technology and AI. These episodes often depicted robots and computers as either benevolent helpers or sinister threats, reflecting societal anxieties about the rapid pace of technological change.

Mainstream publications added to the hype. Magazines like Life and Popular Mechanics ran articles on the potential of AI and automation, often accompanied by futuristic illustrations of robots performing household chores or assisting in complex tasks. These depictions were both exciting and aspirational, promising a future where machines would handle mundane tasks, freeing humans to pursue more meaningful activities.

Notable Figures and Predictions

The 1950s and 1960s were also characterized by the contributions of several key figures in the field of AI, whose work and predictions fueled both optimism and controversy.

Marvin Minsky, a cognitive scientist and co-founder of the MIT Artificial Intelligence Laboratory, was a leading voice in the AI community. Minsky was known for his ambitious vision and confident predictions. In the 1960s, he famously declared that "within a generation... the problem of creating 'artificial intelligence' will substantially be solved." Minsky's optimism was shared by many of his contemporaries, who believed that the rapid advancements in computing would soon lead to the creation of machines capable of human-like intelligence.

John McCarthy, another pivotal figure, coined the term "artificial intelligence" in 1956 and organized the Dartmouth Conference, where AI was formally established as a field of study. McCarthy's work focused on the development of the LISP programming language, which became a fundamental tool for AI research. He envisioned AI systems that could perform tasks such as language understanding, reasoning, and learning from experience.

Herbert A. Simon and Allen Newell, both influential figures in cognitive psychology and computer science, made significant contributions to the field with their work on the Logic Theorist and the General Problem Solver. These early AI programs demonstrated that machines could solve problems that required human intelligence, although within very specific and limited domains.

Examples, Realities, and Fun Facts

The excitement and optimism of the 1950s and 1960s led to several notable projects and developments in AI. However, the realities of the technology often fell short of the high expectations set by its proponents.

One of the earliest AI programs was the Logic Theorist, developed by Allen Newell and Herbert A. Simon in 1955. The Logic Theorist was designed to prove mathematical theorems, and it succeeded in proving 38 of the first 52 theorems in Whitehead and Russell's "Principia Mathematica." This success was a significant milestone, demonstrating that machines could perform tasks that required human intelligence. However, the Logic Theorist also highlighted the limitations of early AI, as it could only operate within the narrow confines of formal logic.

Following the Logic Theorist, Newell and Simon developed the General Problem Solver (GPS), which aimed to solve a broader range of problems by mimicking the human problem-solving process. While the GPS was a significant advancement, it still fell short of the versatile, human-like intelligence envisioned by AI pioneers. The program was limited by the complexity of real-world problems and the computational power of the time.

Another fascinating project from this era was Shakey the Robot, developed at the Stanford Research Institute in the late 1960s. Shakey was one of the first robots capable of navigating and interacting with its environment using a combination of cameras, sensors, and a rudimentary form of AI. Although Shakey's movements were slow and deliberate, it represented a significant step forward in the field of robotics and AI. Shakey could perform tasks such as moving objects and navigating around obstacles, showcasing the potential of AI in physical environments.

Despite these achievements, the gap between the ambitious predictions and the actual capabilities of AI became increasingly apparent. The technology of the time was not yet able to support the grand visions of human-like intelligence and fully autonomous robots. This disparity led to a growing sense of skepticism and, eventually, to the first of the AI winters.

Fun Fact: The Mechanical Turk Hoax

An interesting and often humorous historical footnote in the story of AI is the tale of the Mechanical Turk. In the late 18th century, an inventor named Wolfgang von Kempelen created a machine that purported to be an automated chess-playing device. The Mechanical Turk toured Europe, defeating many human opponents, including notable figures like Benjamin Franklin and Napoleon Bonaparte. However, it was later revealed that the machine was a hoax, with a human chess master hidden inside, controlling the moves. This early deception underscores the long-standing fascination with, and skepticism of, intelligent machines.

The AI Hype Cycle: A Pattern Emerges

The experiences of the 1950s and 1960s reveal a pattern that would repeat throughout the history of

AI: the hype cycle. This cycle begins with a burst of enthusiasm and high expectations, driven by initial successes and bold predictions. As the limitations of the technology become apparent, disappointment and skepticism set in, leading to reduced funding and interest—a period often referred to as an "AI winter." Eventually, new advancements rekindle interest, starting the cycle anew.

This pattern is not unique to AI; it can be seen in many areas of technological innovation. However, AI's cycles of hype and disillusionment are particularly pronounced due to the ambitious nature of its goals—replicating human intelligence is an inherently challenging and complex task.

Conclusion

The 1950s and 1960s were foundational decades for AI, characterized by groundbreaking technological advances and a wave of optimism about the potential of intelligent machines. Pioneers like Alan Turing, Marvin Minsky, John McCarthy, Herbert Simon, and

Allen Newell laid the groundwork for the field, setting ambitious goals and making significant strides. However, the realities of the technology often lagged behind the hype, leading to a growing sense of skepticism and the first AI winter.

Chapter 2: The AI Winters (1970s and 1980s)

Challenges and Setbacks in AI Development

The 1970s and 1980s were marked by a sobering reality check for the field of artificial intelligence. The high expectations set in the previous decades began to meet the hard limitations of technology, theory, and practical application. One of the primary challenges was the computational power required for AI applications. While computers had advanced significantly since the 1950s, they were still far from capable of handling the complex tasks envisioned by AI researchers.

Early AI programs were limited in scope and struggled with problems that required extensive computation and memory. For example, while programs like the Logic Theorist and the General Problem Solver had shown promise in narrow domains, scaling these systems to more complex, real-world problems proved to be a formidable challenge. The algorithms of the time were not efficient enough, and the hardware could not support the vast computational demands of more advanced AI applications.

Another significant challenge was the lack of understanding of human cognition and how to replicate it in machines. AI research at the time was heavily influenced by symbolic AI, which relied on the manipulation of symbols to represent knowledge and reasoning processes. While this approach had some successes, it also had severe limitations, especially in dealing with ambiguous, incomplete, or uncertain information—a common characteristic of real-world problems.

Funding Cuts and Reduced Interest

As the limitations of AI became increasingly apparent, the enthusiasm of the 1950s and 1960s gave way to skepticism and disappointment. This shift in sentiment was reflected in the funding landscape. Government agencies and private investors, previously eager to support AI research, began to pull back their financial support. This period of reduced funding and interest came to be known as the "AI winter."

One of the most significant blows came from the U.S. Department of Defense's Defense Advanced Research Projects Agency (DARPA). In the 1960s, DARPA had been a major funder of AI research, supporting projects at institutions like MIT and Stanford. However, by the mid-1970s, DARPA had become disillusioned with the slow progress and lack of practical results. In 1973, the British government published the "Lighthill Report," which was highly critical of AI research and recommended that funding be significantly reduced. This report had a profound impact, leading to decreased support for AI research in the UK and influencing funding decisions in other countries as well.

The economic climate of the 1970s and 1980s also played a role in the reduction of AI funding. Economic recessions and budget cuts forced many institutions to prioritize more immediate and practical technological developments over speculative research like AI. As a result, many AI projects were shelved, and researchers had to shift their focus to more narrowly defined, achievable goals.

Lessons Learned During the AI Winters

The AI winters were not merely periods of stagnation; they were also times of reflection and learning. Researchers began to reassess the fundamental assumptions underlying their work and sought new approaches to overcoming the challenges they faced.

One of the critical lessons learned was the importance of grounding AI research in practical applications. During the AI winters, researchers who focused on developing systems that could solve specific, well-defined problems found more success than those pursuing grand, generalized AI. For instance, expert systems—AI programs designed to mimic the decision-making abilities of a human expert in a specific domain—emerged as a practical and commercially viable application of AI. Systems like MYCIN, developed in the mid-1970s to diagnose bacterial infections and recommend antibiotics, demonstrated that AI could provide valuable support in specialized areas.

The AI winters also underscored the need for better algorithms and more powerful computing resources. Researchers realized that brute-force approaches, which relied on sheer computational power, were not sustainable. Instead, there was a growing emphasis on developing more efficient algorithms that could solve problems with limited resources. This shift in focus laid the groundwork for future advancements in AI, particularly in the areas of machine learning and neural networks.

Another important development during this period was the increased collaboration between AI researchers and experts in other fields, such as cognitive psychology, neuroscience, and linguistics. This interdisciplinary approach helped to create a more comprehensive understanding of intelligence, both artificial and natural, and provided new insights into how to model cognitive processes in machines.

Examples, Realities, and Fun Facts

Despite the challenges and setbacks, the 1970s and 1980s were not devoid of progress and interesting developments in AI. One notable example was the development of the Stanford Cart, an early autonomous vehicle. The Stanford Cart, created in the late 1960s and improved throughout the 1970s, was a small wheeled robot that could navigate obstacle courses using cameras and a rudimentary form of computer vision. While its movements were slow and deliberate, often taking hours to navigate a simple course, the Stanford Cart represented a significant step forward in the development of autonomous systems.

In the realm of game playing, AI continued to make strides. In 1981, the computer program BELLE, developed by Ken Thompson and Joe Condon, became the first chess machine to achieve a master-level rating. BELLE's success demonstrated the potential of specialized hardware and algorithms tailored to specific tasks, a principle that would become increasingly important in the development of AI.

A fun fact from this era is the story of ELIZA, an early natural language processing program created by Joseph Weizenbaum at MIT in the mid-1960s. ELIZA was designed to simulate a conversation with a psychotherapist using simple pattern-matching techniques. Despite its rudimentary capabilities, many users were surprised by how engaging and lifelike the interactions with ELIZA felt. Weizenbaum himself was astonished by the extent to which people attributed human-like understanding to the program, highlighting the power of even simple AI systems to evoke complex human responses.

The Evolution of AI Techniques

The AI winters prompted researchers to explore new techniques and methodologies that would eventually lead to significant advancements in the field. One of these techniques was machine learning, which focused on developing algorithms that could learn from data and improve their performance over time. While machine learning had been around since the 1950s, it gained renewed attention during the AI winters as researchers sought more effective ways to handle the complexities of real-world problems.

Another area of innovation was neural networks, inspired by the structure and function of the human brain. Although neural networks had been proposed as early as the 1940s, they faced significant challenges in terms of training and scalability. During the AI winters, researchers like Geoffrey Hinton, David Rumelhart, and Ronald Williams made important contributions to the development of backpropagation, a method for training neural networks more effectively. These advancements would later become foundational to the resurgence of AI in the 1990s and beyond.

Conclusion

The 1970s and 1980s were a period of introspection and recalibration for the field of AI. The high expectations and ambitious goals of the previous decades had given way to a more measured and pragmatic approach. Researchers faced significant challenges, including limited computational power, theoretical limitations, and reduced funding. However, these difficulties also provided valuable lessons and spurred the development of new techniques and methodologies that would shape the future of AI.

Chapter 3: The Resurgence of AI in the 1990s

Advances in Computing Power and Algorithms

The 1990s marked a significant turning point for AI, fueled by dramatic advancements in computing power and the development of more sophisticated algorithms. One of the major catalysts for this resurgence was the exponential growth in computational capabilities. Moore's Law, which predicts the doubling of transistors on a microchip approximately every two years, resulted in more powerful and affordable computers. This increased computational power enabled AI researchers to experiment with more complex models and handle larger datasets, essential for training and deploying AI systems.

During this period, the availability of personal computers and advancements in graphics processing units (GPUs) provided the necessary infrastructure for more ambitious AI projects. GPUs, originally designed for rendering graphics, proved to be highly efficient for the parallel processing tasks required by many AI algorithms, particularly in the areas of machine learning and neural networks.

Successful AI Applications

The 1990s saw several high-profile successes in AI, which helped to restore confidence in the field and attract renewed interest and investment. One of the most notable achievements was IBM's Deep Blue, a chess-playing computer that defeated world chess champion Garry Kasparov in 1997. Deep Blue's victory was a landmark event, demonstrating the potential of AI to tackle complex strategic problems and outperform human experts in specific domains.

Deep Blue's success was built on a combination of brute-force computation and sophisticated algorithms. It utilized a massive parallel processing system and employed advanced search techniques to evaluate millions of possible moves per second. The match against Kasparov was closely watched by the global media, highlighting AI's capabilities and sparking widespread public interest.

Another significant AI achievement of the 1990s was the development of autonomous robotic systems.

One notable example was the Mars Pathfinder mission in 1997, which included the Sojourner rover. Sojourner was capable of navigating the Martian surface autonomously, using computer vision and AI algorithms to avoid obstacles and perform scientific experiments. This mission demonstrated the practical applications of AI in challenging and remote environments, showcasing its potential for exploration and data collection.

In the field of natural language processing (NLP), there were also notable advancements. The introduction of statistical methods and probabilistic models transformed NLP, enabling more accurate and robust language understanding. For instance, the development of Hidden Markov Models (HMMs) and their application in speech recognition systems significantly improved the performance and reliability of these systems. This progress laid the foundation for future developments in voice-activated assistants and translation services.

Renewed Interest and Investment

The successes of the 1990s reignited interest and investment in AI research from both the public and private sectors. Governments recognized the strategic importance of AI and began to allocate more funding to support its development. In the United States, agencies like DARPA resumed significant investments in AI, supporting projects that aimed to advance the state of the art in machine learning, robotics, and autonomous systems.

The private sector also played a crucial role in the resurgence of AI. Major technology companies such as IBM, Microsoft, and Google established dedicated AI research labs and invested heavily in developing new AI technologies. These companies recognized the potential of AI to revolutionize various industries and sought to leverage it to gain a competitive edge.

The rise of the internet and the digital economy further fueled interest in AI. The vast amounts of data generated by online activities provided a rich resource for training machine learning models. Companies began to harness this data to develop AI-driven applications, such as recommendation systems, search engines, and personalized advertising. These applications demonstrated the commercial viability of AI and its ability to create significant value.

Examples, Realities, and Fun Facts

The resurgence of AI in the 1990s was marked by several fascinating projects and developments that highlighted both the potential and the challenges of the technology.

One notable example was the development of neural networks and the backpropagation algorithm. Although neural networks had been proposed as early as the 1940s, they faced significant challenges in terms of training and scalability. The backpropagation algorithm, which enabled the efficient training of multi-layer neural networks, was a breakthrough that revitalized interest in this approach. Researchers like Geoffrey Hinton and Yann LeCun made significant contributions to the development and application of neural networks, laying the groundwork for future advancements in deep learning.

Another interesting project from this era was the RoboCup, an international robotics competition that began in 1997. The goal of RoboCup was to advance the state of AI and robotics by challenging teams to develop autonomous robots capable of playing soccer. The competition provided a platform for researchers to test and refine their AI algorithms in a dynamic and unpredictable environment. It also captured the public's imagination, demonstrating the potential of AI in a fun and engaging way.

A fun fact from the 1990s is the story of A.L.I.C.E. (Artificial Linguistic Internet Computer Entity), an early chatbot developed by Richard Wallace in 1995. A.L.I.C.E. used a pattern-matching technique known as Artificial Intelligence Markup Language (AIML) to simulate conversation. Although A.L.I.C.E. was limited in its ability to understand and generate human-like responses, it won several awards in the

Loebner Prize competition, an annual Turing Test contest. A.L.I.C.E.'s success highlighted the progress and ongoing challenges in natural language processing and conversational AI.

The Evolution of AI Techniques

The 1990s also saw the evolution of AI techniques, particularly in the areas of machine learning and data mining. Researchers began to explore new algorithms and approaches that could leverage the increasing availability of data and computational power.

One significant development was the rise of support vector machines (SVMs), a powerful machine learning algorithm introduced by Vladimir Vapnik and his colleagues. SVMs became popular due to their effectiveness in classification tasks and their ability to handle high-dimensional data. They were widely adopted in various applications, from image recognition to bioinformatics.

Another important advancement was the development of ensemble methods, such as bagging and boosting, which combined the predictions of multiple models to improve accuracy and robustness. These techniques proved particularly effective in handling noisy and complex datasets, making them valuable tools for practitioners.

The 1990s also marked the beginning of the era of big data. The proliferation of digital devices and the growth of the internet generated vast amounts of data, providing a rich resource for training machine learning models. Researchers and companies began to recognize the value of data and developed new methods for collecting, storing, and analyzing it. This shift laid the foundation for the data-driven AI approaches that would dominate in the following decades.

Conclusion

The resurgence of AI in the 1990s was a period of significant progress and renewed optimism.

Advances in computing power and algorithms, combined with successful applications and increased investment, reignited interest in the field. High-profile successes like Deep Blue's victory over Garry Kasparov and the autonomous capabilities of the Sojourner rover showcased the potential of AI and captured the public's imagination.

However, this period also underscored the ongoing challenges and limitations of AI. While there were notable achievements, the field still faced significant hurdles in terms of scalability, generalization, and real-world applicability. The lessons learned during the AI winters of the 1970s and 1980s, along with the innovations of the 1990s, set the stage for the next wave of AI development.

Chapter 4: The Rise of Machine Learning and Big Data (2000s)

Introduction of Machine Learning and Its Impact

The turn of the millennium brought about a paradigm shift in the field of AI with the rise of machine learning. Unlike earlier symbolic AI approaches that relied heavily on pre-defined rules and logical reasoning, machine learning focused on developing algorithms that could learn from data and improve their performance over time. This data-driven approach proved to be a game-changer, enabling AI systems to tackle a wider range of complex problems more effectively.

Machine learning's impact was profound across various domains. In the early 2000s, the introduction of more sophisticated algorithms, such as support vector machines (SVMs), decision trees, and ensemble methods, allowed researchers to build models that could handle large, complex datasets. These models were capable of making accurate predictions and uncovering patterns that were previously difficult to detect.

One of the key advantages of machine learning was its ability to generalize from examples, allowing it to perform well on new, unseen data. This capability was particularly valuable in applications such as image and speech recognition, where the variability and complexity of the data made traditional rule-based approaches impractical. Machine learning models, trained on vast amounts of labeled data, could learn to identify objects in images or transcribe spoken words with remarkable accuracy.

Big Data's Role in Fueling AI Advancements

The rise of machine learning was closely intertwined with the explosion of big data. The early 2000s saw an unprecedented growth in the amount of digital data being generated, driven by the proliferation of the internet, social media, and mobile devices. This deluge of data provided the raw material needed to train machine learning models, fueling rapid advancements in the field.

Big data played a crucial role in several high-profile AI applications. For example, search engines like Google leveraged massive datasets to refine their algorithms and deliver more relevant search results. The success of Google's search engine was largely due to its PageRank algorithm, which used data on web link structures to rank the importance of web pages. This data-driven approach set a new standard for the industry and demonstrated the power of big data in enhancing AI performance.

Another area where big data had a transformative impact was in recommendation systems. Companies like Amazon and Netflix used machine learning algorithms to analyze user behavior and preferences, enabling them to provide personalized recommendations for products and content. These recommendation systems became a key driver of user engagement and revenue, highlighting the commercial potential of big data-driven AI applications.

Notable Breakthroughs and Public Reception

The 2000s were marked by several notable breakthroughs in AI, which captured the public's imagination and demonstrated the practical benefits of machine learning and big data.

One of the most significant breakthroughs came in the field of image recognition with the development of convolutional neural networks (CNNs). Yann LeCun, Geoffrey Hinton, and their colleagues made pioneering contributions to the design and training of CNNs, which proved to be exceptionally effective at recognizing objects in images. In 2012, a CNN-based model developed by Hinton's team won the ImageNet Large Scale Visual Recognition Challenge by a wide margin, setting new benchmarks for accuracy and sparking a wave of interest in deep learning.

Speech recognition also saw major advancements during this period. Companies like Google, Microsoft, and IBM developed machine learning models that could transcribe spoken language with high accuracy. These models were integrated into virtual assistants, such as Apple's Siri and Google Assistant, making voice-activated interactions a mainstream feature of smartphones and other devices.

The success of these AI applications had a significant impact on public perception. AI was no longer seen as a futuristic concept; it was becoming an integral part of everyday life. The ability of AI to understand and interact with humans in natural ways—through images and speech—made the technology more accessible and appealing to the general public.

Examples, Realities, and Fun Facts

The 2000s were a decade of rapid innovation and practical deployment of AI technologies, with several interesting projects and developments that highlighted both the potential and the challenges of machine learning and big data.

One notable example was IBM's Watson, an AI system designed to compete on the quiz show "Jeopardy!" In 2011, Watson faced off against two of the show's greatest champions and emerged victorious, demonstrating its ability to process natural language, retrieve information, and reason under time constraints. Watson's success was a major milestone for AI, showcasing the power of machine learning and big data to tackle complex, real-world tasks.

Another significant development was the use of AI in healthcare. Machine learning models were applied to vast datasets of medical records and imaging data to assist in diagnosing diseases, predicting patient outcomes, and personalizing treatment plans. For example, AI systems were developed to analyze mammograms for signs of breast cancer, often matching or surpassing the accuracy of human radiologists. These applications highlighted the potential of AI to improve healthcare delivery and outcomes.

A fun fact from this era is the rise of AI in entertainment. In 2004, a computer program called "Emmy" composed by David Cope, generated original pieces of music in the style of classical composers such as Bach and Mozart. Emmy's compositions were so convincing that they often fooled music experts into thinking they were authentic works by the original composers. This experiment demonstrated the creative potential of AI and sparked debates about the nature of creativity and authorship.

The Evolution of AI Techniques

The 2000s also saw significant advancements in the underlying techniques and methodologies used in AI. One of the most important developments was the resurgence of neural networks, particularly deep learning.

Deep learning, a subset of machine learning, involves training multi-layered neural networks on large datasets. These deep neural networks are capable of learning complex representations of data, making them highly effective for tasks such as image and speech recognition. The development of more efficient algorithms for training deep networks, along with the increased availability of computational power through GPUs, led to a renaissance in neural network research.

Another important advancement was the development of unsupervised learning techniques, which allowed AI systems to learn from data without explicit labels. Algorithms such as clustering and dimensionality reduction became essential tools for analyzing large datasets and uncovering hidden patterns. These techniques were particularly valuable in fields like genomics and social network analysis, where labeled data was often scarce.

Reinforcement learning also gained prominence during this period. This approach, inspired by behavioral psychology, involves training AI agents to make decisions by rewarding them for successful actions and penalizing them for failures. Reinforcement learning proved to be highly effective for tasks that required sequential decision-making, such as game playing and robotic control. Notably, in 2013, DeepMind's reinforcement learning algorithm mastered a range of Atari video games, demonstrating the potential of this approach to learn complex behaviors through trial and error.

Conclusion

The rise of machine learning and big data in the 2000s marked a new era for AI, characterized by significant advancements and practical applications. The shift from symbolic AI to data-driven approaches enabled AI systems to tackle a wider range of complex problems more effectively. The explosion of digital data provided the necessary fuel for training these models, leading to breakthroughs in fields such as image and speech recognition.

High-profile successes, such as IBM's Watson and the development of deep learning, captured the public's imagination and demonstrated the practical benefits of AI. The increased investment and interest from both the public and private sectors further accelerated the progress of AI research and development.

However, this period also highlighted the ongoing challenges and limitations of AI. While machine learning and big data provided powerful tools for analyzing and interpreting information, there were still significant hurdles to overcome in terms of scalability, generalization, and ethical considerations.

Chapter 5: The Current AI Hype (2010s to Present)

Explosive Growth in AI Applications Across Industries

The 2010s witnessed an unprecedented surge in AI applications across various industries, fueled by rapid advancements in machine learning, particularly deep learning, and the availability of vast amounts of data. The combination of these factors led to significant breakthroughs that transformed numerous sectors, from healthcare and finance to transportation and entertainment.

In healthcare, AI-powered systems began to assist in diagnostics, treatment planning, and patient management. For instance, algorithms trained on medical images have been used to detect diseases such as cancer with high accuracy, often matching or surpassing the performance of human radiologists. AI systems also started to predict patient outcomes and recommend personalized treatment plans, improving the quality of care and patient satisfaction.

In finance, AI algorithms have been deployed for fraud detection, risk management, and algorithmic trading. Machine learning models analyze transaction data to identify patterns indicative of fraudulent activity, helping financial institutions prevent fraud in real-time. AI-driven trading algorithms make rapid decisions based on market data, optimizing investment strategies and maximizing returns.

The transportation sector has seen significant advancements with the development of autonomous vehicles. Companies like Tesla, Waymo, and Uber have invested heavily in self-driving technology, aiming to revolutionize the way we travel. These vehicles rely on a combination of sensors, machine learning algorithms, and real-time data to navigate roads, recognize obstacles, and make driving decisions. While fully autonomous vehicles are still in the testing phase, their potential to reduce accidents and improve traffic efficiency is immense.

The Role of Deep Learning and Neural Networks

At the heart of many recent AI advancements is deep learning, a subset of machine learning that involves training multi-layered neural networks on large datasets. Deep learning has been particularly successful in areas such as image and speech recognition, natural language processing, and game playing.

Convolutional neural networks (CNNs) have revolutionized image recognition. These networks are designed to automatically and adaptively learn spatial hierarchies of features from input images. The success of CNNs in the ImageNet competition, where they achieved state-of-the-art performance in object detection and classification, demonstrated their effectiveness and sparked widespread adoption.

Recurrent neural networks (RNNs), and their variants like Long Short-Term Memory (LSTM) networks, have been instrumental in advancing speech recognition and natural language processing (NLP). These networks are designed to handle sequential data, making them well-suited for tasks such as language translation, sentiment analysis, and speech-to-text conversion. Virtual assistants like Apple's Siri, Amazon's Alexa, and Google Assistant rely on these technologies to understand and respond to user queries in natural language.

Generative models, such as Generative Adversarial Networks (GANs) and variational autoencoders (VAEs), have opened new frontiers in AI creativity. GANs, introduced by Ian Goodfellow in 2014, consist of two neural networks—a generator and a discriminator—that compete against each other to create realistic data samples. These models have been used to generate photorealistic images, create art, and even compose music, demonstrating AI's creative potential.

Public and Media Portrayal of AI as a Revolutionary Technology

The media has played a significant role in shaping the public perception of AI as a revolutionary technology. Sensational headlines and dramatic portrayals of AI in movies and television have contributed to a narrative that AI is poised to transform every aspect of our lives. This portrayal, while often exaggerated, has fueled excitement and investment in the field.

One of the most high-profile examples of AI's portrayal in the media is the coverage of AlphaGo, a program developed by DeepMind. In 2016, AlphaGo defeated Lee Sedol, one of the world's top Go players, in a historic match. Go is a complex board game with more possible moves than there are atoms in the universe, making it a formidable challenge for AI. AlphaGo's victory was widely reported in the media and hailed as a major milestone in AI research, showcasing the potential of deep learning and reinforcement learning.

The success of AI in creative fields has also captured public attention. AI-generated art, music, and literature have been featured in exhibitions and concerts, blurring the lines between human and machine creativity. Projects like Google's DeepDream, which uses neural networks to create surreal and dreamlike images, have fascinated and inspired audiences, highlighting the creative possibilities of AI.

However, the media's portrayal of AI as a revolutionary technology has also led to misconceptions and unrealistic expectations. Many people believe that AI is on the verge of achieving human-level intelligence and that robots will soon replace humans in most jobs. This narrative overlooks the significant challenges and limitations that still exist in AI research and development.

Examples, Realities, and Fun Facts

The 2010s and beyond have been marked by several notable AI projects and developments that highlight both the potential and the challenges of AI.

One of the most talked-about AI projects is OpenAI's GPT-3, a state-of-the-art language model capable of generating human-like text. GPT-3, with its 175 billion parameters, can write essays, answer questions, and even generate code. Its ability to produce coherent and contextually relevant text has impressed many, but it has also raised concerns about the ethical implications of such powerful language models, including the potential for misuse in generating fake news or malicious content.

Another significant development is the use of AI in the fight against climate change. AI algorithms are being employed to optimize energy use, improve renewable energy technologies, and monitor environmental changes. For example, machine learning models analyze satellite images to track deforestation and predict the impact of climate change on ecosystems. AI's ability to process vast amounts of data and identify patterns can provide valuable insights for environmental conservation and sustainable development.

A fun fact from this era is the emergence of AI in gaming. In addition to AlphaGo, AI has made strides in other games, such as Dota 2 and StarCraft II. OpenAI's Dota 2 bot, OpenAI Five, competed against professional players and achieved notable victories, demonstrating the potential of reinforcement learning in complex, strategic games. Similarly, DeepMind's AlphaStar achieved Grandmaster level in StarCraft II, showcasing AI's ability to handle real-time strategy and decision-making in dynamic environments.

Challenges and Ethical Considerations

Despite the remarkable progress, the current AI hype has also highlighted several challenges and ethical considerations that need to be addressed.

One of the major challenges is the issue of bias in AI systems. Machine learning models are trained on data that reflects the biases and prejudices present in society. As a result, these biases can be perpetuated and even amplified by AI systems, leading to unfair and discriminatory outcomes. For example, facial recognition systems have been shown to have higher error rates for people of color, raising concerns about their use in law enforcement and surveillance.

Another challenge is the transparency and interpretability of AI models. Many of the most advanced AI systems, such as deep learning models, operate as "black boxes," making it difficult to understand how they arrive at their decisions. This lack of transparency can be problematic in high-stakes applications, such as healthcare and finance, where understanding the rationale behind AI decisions is crucial.

The ethical implications of AI also extend to issues of privacy and security. The widespread use of AI involves the collection and analysis of vast amounts of personal data, raising concerns about data privacy and the potential for misuse. Ensuring that AI systems are secure and that data is handled responsibly is essential to maintaining public trust and preventing harm.

The Importance of Critical Thinking and Skepticism

As AI continues to advance and permeate various aspects of society, it is crucial to maintain a critical perspective and approach the technology with a healthy dose of skepticism.

While AI has the potential to bring about significant benefits, it is essential to recognize that it is not a panacea. AI systems are tools created by humans, and their effectiveness depends on the quality of the data they are trained on, the algorithms used, and the context in which they are applied. Understanding the limitations and potential pitfalls of AI can help prevent overhyped expectations and ensure that the technology is used responsibly.

It is also important to foster an informed and balanced public discourse about AI. Educating people about the capabilities and limitations of AI can help dispel myths and prevent the spread of misinformation. Encouraging critical thinking and skepticism can empower individuals to question and scrutinize AI applications, promoting transparency, accountability, and ethical use of the technology.

Conclusion

The current AI hype, fueled by remarkable advancements in machine learning, deep learning, and big data, has transformed numerous industries and captured the public's imagination. High-profile successes in healthcare, finance, transportation, and entertainment have demonstrated the practical benefits of AI, while the media's portrayal of AI as a revolutionary technology has contributed to its widespread fascination.

However, this period has also highlighted significant challenges and ethical considerations that need to be addressed. Issues of bias, transparency, privacy, and security underscore the importance of developing AI responsibly and ensuring that its benefits are distributed equitably.

Chapter 6: The Realities Behind the Hype

Examination of Genuine AI Advancements Versus Superficial Uses

The current AI hype has led to a proliferation of claims about the technology's capabilities. While there are genuine advancements that have transformed industries and improved lives, many applications of AI are superficial or exaggerated. Distinguishing between real progress and hype is crucial for understanding the true impact of AI and setting realistic expectations.

Genuine AI Advancements

One of the most significant genuine advancements in AI is in healthcare. AI algorithms have made remarkable strides in medical imaging, enabling early and accurate detection of diseases such as cancer. For instance, deep learning models trained on thousands of mammograms can identify signs of breast cancer with accuracy comparable to that of expert radiologists. AI systems are also being used to predict patient outcomes, optimize treatment plans, and assist in drug discovery. These applications have the potential to save lives, reduce healthcare costs, and improve the quality of care.

In the field of natural language processing (NLP), AI has made impressive progress. Language models like GPT-3 can generate human-like text, answer questions, and even write code. These models have numerous applications, including chatbots, virtual assistants, and automated content creation. The ability to understand and generate natural language has opened up new possibilities for human-computer interaction and information retrieval.

Autonomous vehicles are another area where AI has made genuine advancements. Companies like Waymo and Tesla have developed self-driving cars that can navigate complex urban environments, recognize obstacles, and make real-time driving decisions. While fully autonomous vehicles are not yet commonplace, the technology has the potential to reduce traffic accidents, improve mobility, and transform the transportation industry.

Superficial Uses of AI

Despite these genuine advancements, many applications of AI are superficial or overhyped. Companies and startups often claim to use AI to attract investment and attention, even when their use of the technology is minimal or irrelevant. This trend has led to a phenomenon known as "AI washing," where the term "AI" is used as a marketing buzzword rather than a description of actual technology.

For example, some businesses claim to use AI for customer service by deploying simple rule-based chatbots. While these chatbots can handle basic inquiries, they lack the sophistication and understanding of true AI systems. Similarly, many products marketed as "smart" or "AI-powered" rely on basic algorithms or pre-programmed responses rather than genuine machine learning or deep learning.

Another area where AI hype exceeds reality is in predictive analytics. While machine learning models can provide valuable insights and forecasts, their accuracy and reliability depend on the quality and quantity of the data they are trained on. In many cases, businesses overestimate the predictive power of their AI systems, leading to unrealistic expectations and potential failures.

Case Studies of Businesses and Research Projects Misusing AI for Hype

Several case studies illustrate how businesses and research projects have misused AI for hype, leading to negative consequences.

Case Study 1: Theranos

Theranos, a health technology company, claimed to revolutionize blood testing with its proprietary technology. The company touted AI and machine learning as integral to its diagnostic capabilities. However, investigations revealed that the technology was largely ineffective and that the company's claims were exaggerated. The fallout from the scandal resulted in legal action, financial losses, and damaged reputations. Theranos serves as a cautionary tale about the dangers of overhyping AI and the importance of rigorous validation and transparency.

Case Study 2: AI-Powered Recruiting Tools

Several companies have developed AI-powered recruiting tools designed to screen job applicants and predict their suitability for roles. However, these tools have come under scrutiny for perpetuating bias and discrimination. For instance, Amazon abandoned its AI recruiting tool after discovering that it was biased against women. The tool had been trained on resumes submitted over a ten-year period, which reflected historical gender imbalances in the tech industry. This case highlights the risks of relying on biased data and the importance of ethical considerations in AI development.

Case Study 3: Predictive Policing

Predictive policing systems use AI algorithms to analyze crime data and predict where crimes are likely to occur. While these systems are intended to improve law enforcement efficiency, they have been criticized for reinforcing existing biases and disproportionately targeting minority communities. Studies have shown that predictive policing can lead to over-policing in certain areas, exacerbating social inequalities and eroding trust in law enforcement. This case underscores the need for careful evaluation of AI systems and their societal impact.

The Importance of Critical Thinking and Skepticism

In light of these examples, it is clear that critical thinking and skepticism are essential when evaluating AI applications. By asking the right questions and demanding transparency, stakeholders can differentiate between genuine advancements and superficial uses of AI.

Key Questions to Ask

1. **What data is the AI system trained on?** Understanding the source and quality of the training data is crucial. Biased or incomplete data can lead to inaccurate and unfair outcomes.
2. **How does the AI system make decisions?** Transparency in the decision-making process is important for trust and accountability. Stakeholders should seek to understand the algorithms and models used.
3. **What are the limitations and potential biases of the AI system?** Recognizing the limitations and biases of AI systems helps set realistic expectations and prevents overreliance on the technology.
4. **How is the AI system evaluated and validated?** Rigorous testing and validation are essential to ensure the reliability and

effectiveness of AI systems. Stakeholders should look for evidence of thorough evaluation.

5. **What are the ethical implications of using the AI system?** Considering the ethical implications of AI applications helps prevent harm and promotes responsible use of the technology.

Promoting Ethical AI Development

To ensure that AI is developed and used responsibly, it is important to promote ethical AI development practices. This includes:

1. **Bias Mitigation:** Actively working to identify and mitigate biases in AI systems through diverse training data, fairness-aware algorithms, and ongoing monitoring.
2. **Transparency:** Providing clear and accessible information about how AI systems work, including their decision-making processes and limitations.

3. **Accountability:** Establishing mechanisms for holding AI developers and users accountable for the outcomes of their systems, including addressing any harm caused.
4. **Inclusivity:** Involving diverse stakeholders in the development and deployment of AI systems to ensure that different perspectives and needs are considered.
5. **Privacy and Security:** Ensuring that AI systems protect user privacy and data security, and that they comply with relevant regulations and standards.

Conclusion

The realities behind the current AI hype reveal a mixed picture of genuine advancements and superficial uses. While AI has made significant strides in areas such as healthcare, natural language processing, and autonomous vehicles, many applications are overhyped or misused for marketing purposes. Case studies of businesses and research projects that have misused AI highlight the importance of critical thinking, skepticism, and ethical considerations.

Chapter 7: The Human Element

Comparison of AI Capabilities with Human Creativity and Emotions

As AI continues to evolve and integrate into various aspects of our lives, it is crucial to recognize the fundamental differences between AI capabilities and human creativity and emotions. While AI excels at tasks involving data processing, pattern recognition, and automation, it lacks the depth and nuance that characterize human creativity and emotional intelligence.

Human Creativity

Human creativity is a complex and multifaceted phenomenon that involves the generation of novel ideas, solutions, and artistic expressions. Unlike AI, which relies on predefined algorithms and data, human creativity often stems from unique experiences, intuition, and the ability to make connections between seemingly unrelated concepts.

For instance, consider the creativity of artists like Leonardo da Vinci or Vincent van Gogh. Their masterpieces are not just products of technical skill but also of personal expression, cultural context, and emotional depth. AI-generated art, while impressive in its technical execution, often lacks the profound meaning and emotional resonance that characterize human creations.

Moreover, human creativity is not limited to the arts. Scientists and inventors like Albert Einstein and Nikola Tesla demonstrated extraordinary creativity in their ability to conceptualize groundbreaking theories and inventions. Their creative processes involved a deep understanding of their respective fields, the ability to think outside the box, and a willingness to challenge established norms.

Human Emotions

Human emotions play a crucial role in decision-making, interpersonal relationships, and overall well-being. Emotions such as empathy, compassion, and love are integral to the human experience and influence how we interact with others and perceive the world around us.

AI, on the other hand, lacks the capacity for genuine emotions. While AI systems can be programmed to recognize and respond to human emotions to some extent, they do not experience emotions themselves. For example, AI-powered chatbots can detect sentiment in text and provide empathetic responses, but these responses are generated based on patterns in data rather than actual emotional understanding.

The ability to experience and express emotions is also fundamental to human creativity. Many works of art, literature, and music are inspired by the artist's emotions and personal experiences. The emotional depth and authenticity that characterize these works are difficult, if not impossible, for AI to replicate.

The Irreplaceable Qualities of Human Ingenuity

Human ingenuity encompasses a range of qualities that set us apart from AI, including our ability to think critically, solve complex problems, and adapt to new situations. These qualities are essential for innovation and progress and highlight the irreplaceable value of human intelligence.

Critical Thinking and Problem-Solving

Critical thinking involves analyzing information, evaluating evidence, and making reasoned judgments. It requires an understanding of context, the ability to identify biases, and the capacity to consider multiple perspectives. While AI can assist in processing information and identifying patterns, it lacks the nuanced understanding and contextual awareness that underpin human critical thinking.

Problem-solving is another area where human ingenuity shines. Humans can approach problems with creativity, flexibility, and resourcefulness, often finding innovative solutions that go beyond the capabilities of AI. For example, during the Apollo 13 mission, NASA engineers used their ingenuity to develop a makeshift solution to save the astronauts, a feat that relied on human creativity and quick thinking.

Adaptability and Resilience

Humans have a remarkable ability to adapt to changing circumstances and overcome adversity. This adaptability is evident in our response to challenges such as natural disasters, economic crises, and pandemics. Human resilience is driven by a combination of emotional strength, social support, and the capacity to learn from experience.

AI systems, in contrast, are limited by their programming and training data. While they can adapt to some extent through machine learning, their adaptability is constrained by predefined parameters and the quality of the data they are trained on. Human adaptability, fueled by our emotional and cognitive capacities, remains a unique and invaluable asset.

Stories and Quotes from Thought Leaders

Incorporating stories and quotes from thought leaders can provide additional insights into the value of human ingenuity and creativity. Here are a few examples that highlight the importance of these qualities:

Albert Einstein: "Imagination is more important than knowledge. For knowledge is limited, whereas imagination embraces the entire world, stimulating progress, giving birth to evolution."

Einstein's quote underscores the importance of imagination in driving innovation and progress. While AI can process and analyze vast amounts of data, it is human imagination that pushes the boundaries of what is possible.

Steve Jobs: "Creativity is just connecting things. When you ask creative people how they did something, they feel a little guilty because they didn't really do it, they just saw something. It seemed obvious to them after a while."

Jobs' perspective on creativity highlights the ability to make connections between seemingly unrelated ideas, a trait that is uniquely human. This ability to see patterns and opportunities that others may miss is a key driver of innovation.

Maya Angelou: "You can't use up creativity. The more you use, the more you have."

Angelou's quote emphasizes the limitless nature of human creativity. Unlike AI, which relies on existing data and algorithms, human creativity can continually evolve and expand, leading to new ideas and solutions.

The Human-AI Collaboration

While AI and human capabilities are distinct, they can complement each other in powerful ways. By leveraging the strengths of both, we can achieve outcomes that neither could accomplish alone.

Enhancing Human Capabilities

AI can enhance human capabilities by automating routine tasks, providing data-driven insights, and augmenting decision-making processes. For example, AI-powered tools can assist doctors in diagnosing diseases, freeing up time for them to focus on patient care and complex medical decisions. In the creative arts, AI can be used as a tool for artists, providing new mediums and techniques for expression.

Human Oversight and Ethical Considerations

Human oversight is essential to ensure that AI is used responsibly and ethically. While AI can process data and make recommendations, humans must be involved in interpreting the results, making final decisions, and considering the broader ethical implications. This collaborative approach ensures that AI systems are aligned with human values and priorities.

Future Directions and Potential

Looking to the future, the collaboration between humans and AI holds great promise. By combining human creativity, critical thinking, and emotional intelligence with the computational power and data processing capabilities of AI, we can tackle some of the world's most pressing challenges.

Innovations in Healthcare

AI has the potential to revolutionize healthcare by enabling personalized medicine, improving diagnostics, and accelerating drug discovery. However, the successful implementation of these technologies requires the expertise and empathy of healthcare professionals to ensure that patient care remains at the forefront.

Advancements in Education

AI can also transform education by providing personalized learning experiences, identifying gaps in knowledge, and offering real-time feedback. Educators can use AI tools to enhance their teaching methods, while also ensuring that the human aspects of mentorship and support remain integral to the learning process.

Sustainable Development

AI can play a critical role in addressing environmental challenges by optimizing resource use, monitoring ecosystems, and predicting climate change impacts. Human ingenuity is needed to develop and implement sustainable solutions that balance technological advancements with environmental preservation.

Conclusion

The human element remains irreplaceable in the age of AI. While AI has made significant strides and offers powerful tools for various applications, it is human creativity, critical thinking, and emotional intelligence that drive true innovation and progress.

Chapter 8: The Future of AI: Beyond the Hype

Predictions for AI's Trajectory in the Coming Decades

As we look towards the future, the trajectory of AI is both promising and challenging. Predicting the exact path of AI's development is difficult, but several trends and potential directions stand out.

AI Integration into Everyday Life

AI is likely to become increasingly integrated into everyday life, extending its influence across various domains. Smart home devices, personal assistants, and wearable technologies will become more sophisticated, offering enhanced convenience and personalized experiences. For instance, AI-driven health monitors could continuously track vital signs and provide real-time health insights, helping individuals manage their well-being more proactively.

Advancements in Autonomous Systems

Autonomous systems, including self-driving cars, drones, and robotic assistants, will continue to evolve. As these technologies become more reliable and widespread, they have the potential to transform industries such as transportation, logistics, and healthcare. Autonomous vehicles could reduce traffic accidents and congestion, while drones might revolutionize delivery services and disaster response efforts.

AI in Scientific Research and Discovery

AI will play an increasingly crucial role in scientific research and discovery. Machine learning algorithms can analyze vast datasets, uncovering patterns and insights that might elude human researchers. AI-driven simulations and models could accelerate breakthroughs in fields such as physics, chemistry, and biology. For example, AI has already been used to predict protein structures, a critical step in understanding diseases and developing new treatments.

Ethical AI and Responsible Development

As AI continues to advance, ethical considerations will become even more important. Ensuring that AI systems are developed and deployed responsibly will be essential to prevent harm and promote fairness. This includes addressing issues such as bias, transparency, privacy, and accountability. Policymakers, researchers, and industry leaders will need to collaborate to establish guidelines and standards for ethical AI development.

The Potential for AI to Mature and Integrate Meaningfully

For AI to realize its full potential, it must mature and integrate meaningfully into society. This involves addressing technical challenges, fostering public trust, and ensuring that AI systems complement rather than replace human capabilities.

Technical Challenges and Innovations

Several technical challenges must be overcome for AI to mature. These include improving the robustness and reliability of AI systems, enhancing their ability to generalize from limited data, and developing better techniques for explainability and interpretability. Innovations in these areas will be crucial for building AI systems that are both effective and trustworthy.

Fostering Public Trust

Public trust in AI is essential for its successful integration into society. This requires transparent communication about the capabilities and limitations of AI, as well as clear guidelines for its ethical use. Engaging with diverse stakeholders, including the public, industry, and academia, can help build a consensus on the responsible development and deployment of AI technologies.

Complementing Human Capabilities

AI should be seen as a tool that complements human capabilities rather than replacing them. By leveraging the strengths of both AI and human intelligence, we can achieve outcomes that neither could accomplish alone. This collaborative approach will be key in areas such as healthcare, education, and creative industries, where the human touch remains irreplaceable.

Ethical Considerations and the Importance of Responsible AI Development

As AI becomes more powerful and pervasive, ethical considerations will be paramount. Ensuring that AI is developed and used responsibly will be essential to maximize its benefits and minimize potential harms.

Addressing Bias and Fairness

AI systems are only as good as the data they are trained on. If the training data contains biases, the AI system will likely perpetuate those biases. It is crucial to develop techniques for identifying and mitigating bias in AI models. This includes using diverse and representative datasets, as well as implementing fairness-aware algorithms.

Ensuring Transparency and Accountability

Transparency is vital for building trust in AI systems. This involves making the decision-making processes of AI systems understandable and accessible. Explainable AI (XAI) techniques can help by providing insights into how AI models arrive at their decisions. Additionally, accountability mechanisms must be established to hold developers and users of AI systems responsible for their actions and the outcomes of their technologies.

Protecting Privacy and Data Security

The widespread use of AI involves the collection and analysis of vast amounts of personal data. Protecting the privacy and security of this data is essential to prevent misuse and ensure that individuals' rights are respected. This requires robust data protection measures, compliance with relevant regulations, and the development of privacy-preserving AI techniques.

Promoting Ethical Guidelines and Standards

Developing and promoting ethical guidelines and standards for AI is a collective responsibility. Policymakers, industry leaders, researchers, and civil society organizations must work together to establish norms and frameworks for responsible AI development. These guidelines should address issues such as bias, transparency, privacy, accountability, and the broader societal impact of AI.

Conclusion

The future of AI holds tremendous potential, but realizing this potential requires careful consideration of both the opportunities and challenges that lie ahead. As AI continues to advance, it will become increasingly integrated into everyday life, transforming industries and enhancing human capabilities. However, this progress must be guided by ethical principles and a commitment to responsible development.

Chapter 9: Will AI Take Your Job?

The Impact of AI on the Workforce

As AI continues to advance and integrate into various industries, a pressing question arises: Will AI take your job? The impact of AI on the workforce is a complex and multifaceted issue that involves considerations of automation, job displacement, and the creation of new opportunities. Understanding these dynamics is crucial for preparing for the future of work.

Automation and Job Displacement

AI and automation have the potential to significantly alter the job landscape. Many routine and repetitive tasks, which are common in industries such as manufacturing, logistics, and customer service, are prime candidates for automation. For instance, assembly line work, data entry, and basic customer inquiries can be efficiently handled by AI-powered systems and robots.

Examples of Job Automation

1. **Manufacturing**: In manufacturing, AI-driven robots and automated systems can perform tasks such as welding, painting, and assembly with high precision and efficiency. This reduces the need for human labor in these roles and increases production speed and consistency.
2. **Logistics**: In the logistics sector, AI is used to optimize supply chain management, predict demand, and automate warehouse operations. Autonomous vehicles and drones are being developed to handle delivery tasks, potentially reducing the demand for human drivers and delivery personnel.
3. **Customer Service**: AI chatbots and virtual assistants are increasingly being used to handle routine customer service inquiries. These systems can answer common

questions, process orders, and provide technical support, reducing the need for human customer service representatives.

The Creation of New Opportunities

While AI will undoubtedly displace some jobs, it will also create new opportunities and demand for different skill sets. The key to navigating this transition is understanding where these opportunities lie and how to adapt.

Emerging Job Roles

1. **AI Development and Maintenance**: As AI technology proliferates, there will be a growing demand for professionals who can develop, implement, and maintain AI systems. This includes roles such as AI engineers, data scientists, machine learning specialists, and AI ethics consultants.
2. **Human-AI Collaboration**: Many jobs will evolve to involve collaboration between humans and AI systems. For example,

doctors will use AI to assist in diagnostics and treatment planning, while journalists may use AI to help with data analysis and content creation. These roles will require individuals to have both domain-specific knowledge and an understanding of how to leverage AI tools effectively.

3. **Creative and Strategic Roles**: Jobs that require creativity, strategic thinking, and complex problem-solving are less likely to be automated. Roles in fields such as marketing, design, research, and management will continue to be in demand. AI can augment these roles by providing insights and data-driven recommendations, but human ingenuity will remain essential.

The Importance of Upskilling and Reskilling

Adapting to the changing job landscape will require a focus on upskilling and reskilling the workforce. Continuous learning and professional development will be crucial for individuals to remain competitive in the job market.

Educational Initiatives

1. **Technical Skills Training**: Programs that teach technical skills such as coding, data analysis, and machine learning will be vital. These skills will enable individuals to work with AI technologies and take on roles in the growing tech sector.
2. **Soft Skills Development**: Equally important will be the development of soft skills such as critical thinking, creativity, emotional intelligence, and adaptability. These skills will help individuals navigate the complexities of human-AI collaboration and excel in roles that require human intuition and empathy.
3. **Lifelong Learning**: Encouraging a culture of lifelong learning will be essential. Educational institutions, employers, and governments must work together to provide

accessible and flexible learning opportunities that allow individuals to continually update their skills and knowledge.

Case Studies of AI and Job Transformation

Examining case studies can provide valuable insights into how AI is transforming jobs and industries.

Case Study 1: The Banking Sector

In the banking sector, AI has been used to automate routine tasks such as processing loan applications, detecting fraudulent transactions, and providing customer support. While this has reduced the need for certain clerical roles, it has also created opportunities for roles in AI development, cybersecurity, and data analysis. Additionally, bank employees are now more focused on providing personalized financial advice and developing customer relationships, tasks that benefit from human empathy and understanding.

Case Study 2: The Healthcare Industry

In healthcare, AI systems assist in diagnostics, treatment planning, and patient monitoring. For example, AI algorithms can analyze medical images to detect diseases at an early stage, improving patient outcomes. While AI can handle some diagnostic tasks, it does not replace doctors and nurses. Instead, it augments their capabilities, allowing them to spend more time on patient care and complex medical decision-making.

Case Study 3: The Retail Industry

In retail, AI is used to optimize inventory management, personalize marketing, and improve customer service. Automated checkout systems and AI-driven recommendation engines are becoming more common. While this reduces the need for some retail workers, it also creates new roles in AI system management, digital marketing, and customer experience design.

Addressing the Societal Impact of AI on Jobs

The widespread adoption of AI will have significant societal implications. Addressing these impacts requires a proactive approach from governments, businesses, and educational institutions.

Government Policies

1. **Support for Workers**: Governments can provide support for workers displaced by AI through unemployment benefits, retraining programs, and job placement services. Policies that promote job creation in emerging industries can also help mitigate the impact of job displacement.
2. **Education and Training**: Investing in education and training programs that prepare individuals for the jobs of the future is essential. This includes funding for technical education, lifelong learning initiatives, and partnerships with industry to align training programs with market needs.
3. **Regulation and Standards**: Developing regulations and standards for the ethical use

of AI can help ensure that the technology is deployed in a way that benefits society. This includes addressing issues such as bias, transparency, and accountability in AI systems.

Business Strategies

1. **Workforce Transition Planning**: Businesses can develop strategies for transitioning their workforce to adapt to AI. This includes providing retraining and upskilling opportunities, fostering a culture of continuous learning, and creating new roles that leverage AI technologies.
2. **Collaboration with Educational Institutions**: Businesses can collaborate with educational institutions to design training programs that meet industry needs. This can help ensure that graduates are equipped with the skills required for the evolving job market.

3. **Ethical AI Deployment**: Companies should commit to the ethical deployment of AI, ensuring that their AI systems are fair, transparent, and accountable. This includes regularly auditing AI systems for bias and implementing measures to mitigate any negative impacts on employees and customers.

Educational Institutions

1. **Curriculum Development**: Educational institutions can develop curricula that incorporate both technical skills and soft skills, preparing students for the future job market. This includes offering courses in AI, machine learning, data science, and ethics, as well as fostering creativity, critical thinking, and emotional intelligence.
2. **Lifelong Learning Opportunities**: Providing lifelong learning opportunities, such as online courses, workshops, and certification programs, can help individuals

continuously update their skills and stay competitive in the job market.

3. **Industry Partnerships**: Collaborating with industry partners can help educational institutions stay current with market trends and ensure that their programs are relevant and aligned with industry needs.

Conclusion

The question of whether AI will take your job is complex and multifaceted. While AI will undoubtedly lead to job displacement in some sectors, it will also create new opportunities and demand for different skill sets. The key to navigating this transition is understanding where these opportunities lie and how to adapt.

Chapter 10: Artificial Illusion

The Illusion of Groundbreaking Innovations

In the current landscape of technological advancement, the term "AI" is often thrown around with great enthusiasm and little substance. During conversations with CEOs of startups claiming to use AI for groundbreaking innovations, it becomes evident that many are engaging in what can be termed "Artificial Illusion." When asked about the specific models they use, the reasoning behind their choices, and the particular applications, the responses are often vague, encapsulated in broad and non-specific references to AI.

This phenomenon is reminiscent of the saying, "Fake it till you make it," but with a modern twist: "Artificial Illusion." As Mark Twain aptly remarked, "It's easier to fool people than to convince them that they have been fooled." This statement holds particularly true in the realm of AI hype, where the mere mention of AI can lend credibility and attract resources, even when the actual implementation is minimal or superficial.

The Dangers of Artificial Illusion

The widespread enthusiasm for AI has created an environment where merely mentioning the technology can enhance a company's image and attract investment. This trend, however, is fraught with risks. By overemphasizing AI without substantial implementation, companies dilute the genuine potential of AI and risk disillusioning stakeholders who may eventually see through the facade.

Dilution of Genuine Potential

When companies engage in Artificial Illusion, they shift focus away from meaningful innovation to superficial enhancements aimed at capitalizing on the AI hype. This not only misleads investors and customers but also undermines the credibility of truly innovative AI projects. The result is a marketplace cluttered with overstated claims and underdelivered promises, which can stymie the progress of authentic advancements.

Risk of Disillusionment

The disillusionment of stakeholders, including investors, customers, and employees, can have far-reaching consequences. When the promised AI capabilities fail to materialize, trust erodes, and skepticism increases. This disillusionment can lead to a reduction in investment, a decrease in customer loyalty, and a loss of morale among employees who may feel they are part of a deceptive endeavor.

True Innovation Requires Substance

True innovation in AI demands more than just buzzwords; it requires substance, transparency, and a commitment to solving real problems. Companies that genuinely leverage AI for innovation do so by developing robust models, rigorously validating their effectiveness, and transparently communicating their capabilities and limitations.

Case Studies in Artificial Illusion

Case Study 1: The Chatbot Hype

Several companies have claimed to revolutionize customer service with AI-powered chatbots. However, upon closer examination, many of these chatbots are based on simple rule-based systems rather than sophisticated AI. These systems can handle basic inquiries but fail to manage complex interactions or learn from past interactions. This discrepancy between claim and capability often leads to customer frustration and tarnishes the company's reputation.

Case Study 2: Predictive Analytics

In the realm of predictive analytics, some businesses tout their AI systems' ability to forecast market trends or consumer behavior with high precision. However, many of these claims are based on rudimentary statistical models rather than advanced machine learning techniques. The overstatement of their capabilities can lead to misguided business strategies and financial losses.

The Role of Skepticism and Critical Thinking

In the face of Artificial Illusion, skepticism and critical thinking become invaluable tools. It is essential to question the claims made about AI and seek evidence of genuine innovation. As consumers, investors, and professionals, we must learn to distinguish between true advancements and mere marketing ploys.

Questions to Ask

1. **What specific AI models and algorithms are being used?** Understanding the technical details can provide insight into the depth of the AI implementation.
2. **How is the AI system trained and validated?** Rigorous training and validation processes are indicators of a robust AI system.
3. **What problem is the AI system solving, and how effectively does it do so?** Clarity on the problem and measurable outcomes can differentiate genuine innovation from hype.

4. **Is there transparency in how the AI system operates?** Transparency in AI operations fosters trust and accountability.

Historical Wisdom on Modern Challenges

If Shakespeare were alive today, he might quip, "To AI, or not to AI, that is the question." The answer lies in recognizing that while AI has its place, it is not the end-all and be-all. We should remember Albert Einstein's wise words: "Imagination is more important than knowledge." AI lacks the imagination and creativity that humans bring to the table, and it is this human element that drives true innovation.

The real secret to success isn't in the latest tech jargon but in honest, hard work and genuine innovation. As George Bernard Shaw aptly put it, "Beware of false knowledge; it is more dangerous than ignorance." This cautionary note is particularly relevant in the context of AI, where the allure of Artificial Illusion can obscure the path to real progress.

Conclusion

The concept of Artificial Illusion serves as a reminder that true innovation requires more than just invoking the latest buzzwords. It demands substance, transparency, and a commitment to solving real problems.

Conclusion

As we conclude our book of artificial intelligence, it's evident that AI is a powerful and transformative technology with the potential to reshape many aspects of our lives. From its early beginnings in the 1950s to the complex, data-driven systems of today, AI has evolved significantly, experiencing cycles of hype and disillusionment, breakthroughs and setbacks. This evolution highlights both the extraordinary capabilities of AI and the limitations that remain.

AI's dual nature—its ability to solve complex problems and its susceptibility to overhyped claims—presents a paradox that must be navigated carefully. On one hand, genuine AI advancements have revolutionized fields such as healthcare, finance, transportation, and scientific research, offering unprecedented opportunities for innovation and progress. On the other hand, the phenomenon of Artificial Illusion serves as a reminder of the dangers of superficial applications and exaggerated promises.

Human ingenuity, creativity, and emotional intelligence are irreplaceable qualities that AI lacks. While AI can process vast amounts of data and perform specific tasks with remarkable efficiency, it does not possess the depth of understanding, empathy, and imaginative capacity that define human intelligence. The true potential of AI lies not in replacing humans but in augmenting our abilities and enhancing our decision-making processes.

As AI continues to advance, ethical considerations and responsible development must remain at the forefront. Addressing issues such as bias, transparency, privacy, and accountability is essential to ensure that AI benefits society as a whole. Policymakers, industry leaders, researchers, and the public must collaborate to establish guidelines and standards that promote fairness and prevent harm.

The future of AI holds both promise and challenges. It is essential to prepare for the changes that AI will bring to the workforce, education, and daily life. Fostering a culture of lifelong learning, investing in upskilling and reskilling, and promoting ethical AI development will help navigate the complexities of the AI landscape and create a more equitable and prosperous future.

Innovation is the driving force behind progress, but it must be embraced with caution and critical thinking. The allure of new technology should not blind us to the importance of rigorous validation, transparency, and ethical considerations. Asking the right questions and demanding evidence of true innovation will ensure that AI lives up to its potential and serves the greater good.

The road ahead for AI is both exciting and uncertain. As we continue to explore the possibilities and address the challenges, maintaining a balanced perspective is crucial. AI is a tool—a powerful one—but it is the human element that will ultimately determine its impact on society. Leveraging our unique qualities and collaborating with AI can unlock new opportunities and create a better future for all.

The development of AI is far from over. Lessons learned from its history, insights gained from its present, and the potential envisioned for its future will guide us as we move forward.

www.ingramcontent.com/pod-product-compliance
Lightning Source LLC
Chambersburg PA
CBHW031446210526
45464CB00005B/2347